# 5-MeO-DMT Source and extraction
# A Complete Guide to Finding and extracting 5-MeO-DMT

**Trevor Turner**

Copyright@2024 Trevor Turner all right reserved. No part of this publication should be or transmitted in any form or means without prior written permission from the copyright holder

Chapter one

   5meo dmt Source and extraction

Chapter Two

   What Makes 5-MeO-DMT Different from Other Types of DMT

Chapter Three

   Some plants that contain 5-MeO-DMT

Chapter four

   Animal sources

Chapter five

   Tools You Need to Get 5-MeO-DMT

Chapter six

   Step by step extraction from plant

Chapter seven

   Extraction of 5-MeO-DMT from toad venom

Chapter eight

   Safe ways to handle 5-MeO-DMT

## Chapter one

## 5meo dmt Source and extraction

**What is 5-MeO-DMT?**

5-Methoxy-N,N-Dimethyltryptamine, or 5-MeO-DMT, is a strong mind-altering chemical that is in the tryptamine family. This chemical is natural and can be found in many plants and the poison of the Colorado River Toad (Bufo alvarius). Some important things about 5-MeO-DMT are these:

**The chemicals structure and properties**

• **chemicals structure:** 5-MeO-DMT is made up of a tryptamine backbone and a methoxy group (-OCH3) connected to the fifth carbon of the indole ring. The chemical formula is $C_{13}H_{18}N_2O$.

- **Solubility:** It dissolves quickly in organic fluids but not so much in water.

### Effects and Experiences

- **Effects Start and Last:** 5-MeO-DMT effects start quickly, usually within seconds to minutes when burned or vaporized, and can last for anywhere from 15 to 45 minutes.
- **Psychoactive Effects:** People who use it say they have deep, intense experiences that they often call divine or mystical. A sense of unity, the loss of ego,

changes in how time and place are perceived, and deep emotional or spiritual insights are all common effects.

- **Physical Effects:** An elevated heart rate, changes in blood pressure, and physical feelings like burning or warmth are some of these effects.

**Sources from nature**

- **Plants:** 5-MeO-DMT can be found in many types of plants, such as the yopo Anadenanthera peregrina and the Virola species.

**Animal:** People usually get 5-MeO-DMT from animals. The Colorado River Toad (Bufo alvarius) is the most famous animal source.

**History and cultural Significance**

• For hundreds of years, native tribes in South America have used plants that contain 5-MeO-DMT for healing and shamanic purposes.

• People are more interested in the substance now because it might be useful in medicine and has strong psychoactive effects.

## Chapter Two

## What Makes 5-MeO-DMT Different from Other Types of DMT

5-MeO-DMT, N,N-DMT, and 4-AcO-DMT are all psychoactive compounds that belong to the tryptamine class. However, their chemical structures, effects, natural sources, and legal situations are all different. Here's what makes them different:

**The Chemicals Structure**

**• 5-MeO-DMT (5-Methoxy-N,N-Dimethyltryptamine)**

A methoxy group (-OCH3) is linked to the fifth carbon of the indole ring, giving it its shape.

- **N,N-DMT (N,N-Dimethyltryptamine)**

The tryptamine structure is simple, with no extra functional groups connected to the indole ring.

- **4-AcO-DMT (4-Acetoxy-N,N-Dimethyltryptamine)**

On the fourth carbon of the indole ring, there is an acetoxy group (-OCOCH3) attached.

**What Happens and What We Feel**

- **5-MeO-DMT**
- Effects: Strong and quick to start, often said to be stronger and more powerful than N,N-DMT. Some of the things that can happen are ego dissolution, a

sense of unity, and deep spiritual or mystical understanding.
- Length: Not long, usually 15 to 45 minutes.

- **N,N-DMT**
- Effects: Bright, colorful, and often geometric visual images, along with a strong feeling of entering other worlds or dimensions. Deep spiritual and introspective moments are also possible.

The length is short, usually between 5 and 30 minutes.

- **4-AcO-DMT**
- the effects are a lot like psilocybin (found in magic mushrooms), since the body turns it into psilocin. Visual and

audio hallucinations, changes in time and reality, and deep emotional and mental insights are some of the things that people can experience.

- Lasts longer than 5-MeO-DMT and N,N-DMT, usually 3 to 6 hours.

## Natural Sources

### • 5-MeO-DMT

This poison is found in many plants, such as Anadenanthera peregrina and Virola species, and in the venom of the Colorado River Toad (Bufo alvarius).

### • N,N-DMT

It is found in a lot of different plant and animal species, like Psychotria viridis and Mimosa hostilis.

- **4-AcO-DMT**
- Synthetic compound: It doesn't happen naturally, but it is very similar to psilocybin, which is found in some types of magic mushrooms.

**How to Use and Why**

- **5-MeO-DMT**

Because it has strong spiritual benefits, it is often used in ceremonies or as medicine. Because it has strong affects, it needs to be carefully planned and used in a controlled environment.

- **N,N-DMT**

It is often used in a variety of traditional practices (like Ayahuasca events) and for fun because it has strong hallucinogenic effects.

- **4-AcO-DMT**

usually used for fun or medical reasons to get effects similar to psilocybin. Some users like it better than psilocybin because it has more predictable effects and is allowed in some places

**Note:** 5-MeO-DMT, N,N-DMT, and 4-AcO-DMT all come from the same tryptamine base, but their effects, duration, natural sources, and legal standing are all very different. 5-MeO-DMT is known for its strong effects that start quickly. N,N-DMT causes vivid and complex visual hallucinations.

And 4-AcO-DMT has effects that last longer, like psilocybin.

## Chapter Three

## Some plants that contain 5-MeO-DMT

There are several plants that contain 5-MeO-DMT, and they are all found in different places and have different societal meanings. These plants are well-known for having 5-MeO-DMT in them:

**1. Anadenanthera peregrina (Yopo and Cohoba have).**

• Fabaceae family

• **where it lives:** It is native to South America and the Caribbean.

• **Traditional Use:** The seeds of this tree are used to make a smoke called yopo or cohoba by native people, especially those who live in the Amazon Basin. To make a powder that is blown into the user's nose, the seeds are toasted, ground, and mixed with other things.

• **Active Compounds:** 5-MeO-DMT and bufotenine can be found in the seeds.

## 2. species of Virola (Espina and Parica)

• **Plant family:** Myristicaceae

• **Geographical Distribution:** It lives in the Amazon jungle in places like Colombia, Peru, and Brazil.

• **In traditional medicine,** the oil from the bark of Virola trees is used to make a snuff that makes people feel high. The resin is gathered, heated, and then mixed with ash or other things before it is used.

- **Active Compounds:** 5-MeO-DMT, DMT, and other tryptamines are found in the resin.

### 3. The Diplopterys cabrerana (Chagropanga and Chaliponga) species

- **Type of plant:** Malpighiaceae
- **Geographical Distribution:** They live in the Amazon jungle in places like Ecuador, Peru, and Colombia.
- **The traditional use** of this plant is to add it to Ayahuasca drinks. It is mixed with Banisteriopsis caapi to make it

more psychoactive.

- **Active Ingredients:** 5-MeO-DMT, DMT, and other chemicals are present.

### 4. The gangeticum desmodium

- Fabaceae family
- **where it lives:** It lives in South Asia, in places like India, Sri Lanka, and the Himalayas.
- **Traditional Use**: It is used for many things in Ayurvedic health.
- **Active Compounds:** It has 5-MeO-DMT in it, but not as much as certain other plants.

### 5. Phalaris species (Canary grass).

- Poaceae family
- **Geographical Distribution:** Different kinds can be found all

over the world, mostly in temperate areas.

• **Traditional Use**: Some species of Phalaris grass have been found to contain psychoactive tryptamines, even though they haven't generally been used as a drug.

**Chemicals That Work:** Some species, like Phalaris arundinacea and Phalaris brachystachys, have 5-MeO-DMT, DMT, and other similar chemicals.

## 6. Mimosa tenuiflora (Jurema and Tepezcohuite)

• Fabaceae family

• **where it lives:** It is native to Brazil and other parts of South America.

- **Traditional Use:** It is used in traditional Brazilian medicines like Jurema and Ajucá, and it can also be used to make people feel high.
- **Active Compounds:** mostly DMT, but smaller amounts of 5-MeO-DMT may also be present.

These plants have very different traditional uses, where they grow, and how much 5-MeO-DMT they contain. They have been very important to the spiritual and cultural traditions of native people, especially in South America. When these plants are taken out of their natural environments or used, legal issues should be taken into account.

## Chapter four

## Animal sources

The Colorado River Toad (Bufo alvarius), which is also called the Sonoran Desert Toad, is the main animal source of 5-MeO-DMT. Here are the specifics:

**The Colorado River Toad, or Bufo alvarius,**

- **Geographical Distribution:** In the southwestern United States and northern Mexico, especially in the Sonoran Desert, you can find this toad.
- **Appearance**: It looks like a big toad, about 4 to 7 inches long, with smooth, greenish-brown skin.

- **Excretion:** The toad's parotoid glands, which are behind its eyes and on its legs, make venom. This poison has a lot of chemicals that make you feel high, like 5-MeO-DMT and bufotenine.
- **The process of extraction:**

●Usually, the venom is taken out by carefully pressing on the parotoid glands. To do this, the glands must be slightly squeezed to release the venom onto a smooth surface, like glass.

●After being left to dry, the venom turns into a crystallized material that can be scraped off and saved for later use.

●When burned and breathed in, the 5-MeO-DMT in the dried

venom has strong psychoactive effects.

**Ethics Things to Think About**

- **Sustainability:** Toad populations should stay healthy; collecting poison should be done in an honest way that doesn't hurt the toads.
- **Conservation:** The Colorado River Toad is in danger because its environment is being destroyed and it is being overharvested. To protect this species and its habitat, conservation measures are very important.

**Other Options:** Synthetic 5-MeO-DMT can be used instead of toad venom to get the same high results without hurting animals.

**Locating Sources**

**1.How to Identify Plants in the Field**

- **Research and planning:**
Learn about the unique traits of the plants that contain 5-MeO-DMT, such as how they grow, the shape of their leaves, and where they usually live.

- **Field Guides and Local Experts:** To get a correct identification, use botanical field guides or talk to local experts and native groups.

- **Ethical Harvesting:** Make sure that gathering is done in a way that doesn't hurt the environment or local ecosystems.

## 2.How to Find and Handle Toads

- **Finding Toads:** Toads are most busy during the wet season or right after it rains, so look for them in their natural habitats during those times.
- **Ethical Treatment:** Be careful not to hurt or stress toads when handling them. After getting the venom, put them back where they belong.

# Chapter five

## Tools You Need to Get 5-MeO-DMT

**Collection of Plants**

**1. Identification guide and field books**

• Botanical Field Guides: To correctly name the plants that have 5-MeO-DMT in them.

• Local Flora Books: These are specific to the area where you are collecting plants.

**2. Tools for Harvesting**

• For cleanly cutting leaves, stems, or seed pods, pruners or secateurs are best.

• A sharp knife or machete is useful for cutting through bigger twigs or bark, especially on trees like the Virola species.

• A shovel or rake for pulling small plants out of the ground if needed.

## 3. Containers for collection and storage

Mesh or cloth bags are used to collect plant parts like seeds and leaves so they can breathe and keep mold from growing.

• Plastic bags or containers: For short-term keeping, especially if you're worried about moisture.

## 4. Equipment for drying and processing

• Drying Rack or Screen: For putting plants out to dry on their own.

• Dehydrator: Drying that can be managed better and removes moisture more evenly.

Large trays or sheets: These are used to spread out bark or leaves while they dry.

## 5. Wear gloves and safety gear

• Gardening gloves: to keep your hands safe from sharp tools and plant materials that could be irritating.

• Protective clothing, like long arms and pants to keep you from getting scratches and plant sap on your skin.

## 6. Field Gear

• Backpack: to carry tools and things you've gathered.

- A notebook and pen: to write down information about where plants are and what they look like.
- A GPS device or smart phone: to find collection spots and make maps of them.

**For the collection of Toad Venom**

**1. Identification guide and field books**

- To correctly name the Colorado River Toad (Bufo alvarius), use herpetology guides.

**2. Tools for Collection**

- Gloves: Thick, safe gloves to keep your hands and the toad safe while you handle it.
- Plastic or glass sheets: Put the

sheet under the toad's parotoid glands while it milks to get the venom.

• Soft brushes: These are used to carefully remove the dried poison from the glass or plastic sheet.

### 3. Containers for storage

• Containers that don't let air in: These are used to keep dried venom crystals free of wetness and other harmful substances.

• Drying Packs: To put in storage cases to keep the poison dry.

### 4. Field Gear

• Headlamp or flashlight: to find toads at night when they are busy.

• A backpack to carry containers and tools for gathering.

• A notebook and pen: Keep track of where the toads are and what their situations are like.

## Extraction method for 5-MeO-DMT

It takes several steps to separate and clean 5-MeO-DMT from plant matter or toad poison that comes from natural sources. The steps are a little different based on the source. We'll talk about how to get toad venom and plant sources extracted here.

## Chapter six

## Step by step extraction from plant

getting 5-MeO-DMT from plants requires a few important steps to separate the chemical from the plant matter. Here is a step-by-step guide:

**Material needed:**

• Plant parts that have 5-MeO-DMT in them, like Anadenanthera peregrina seeds, Virola species bark, and Diplopterys cabrerana leaves.

• A mix of acids, like hydrochloric acid or acetic acid

• A base solution, like sodium hydroxide

- Solvent that isn't polar, like naphtha or heptane
- Water with a filter
- Glassware, like flasks and beakers
- Magnetic stirrer or a stirring rod
- A separator pipe
- Paper coffee cups or filter paper
- pH gauge or pH strips
- Dish for evaporation
- A heat generating source, like a stove or hot plate
- Safety gear (goggles, gloves, and a smoke hood)

**How to do it:**

**1. Getting Plant Material Ready**

- **Drying:** Make sure the plant matter is completely dry to keep mold away and make grinding easy.

● **Grind:** Use a grinder or a mortar and pestle to turn the dried plant matter into a fine powder. This gives the extraction process more surface area.

**2. Acidic Extraction**

● **mixing:** put the powdered plant material in a glass jar and add the acidic solution (such as hydrochloric acid or diluted acetic acid). The acid changes 5-MeO-DMT into a salt that can dissolve in water.

● **Stirring:** Make sure all the plant parts are covered in the acidic solution by stirring the mixture well.

- **Soaking:** Let the mixture soak for a few hours to overnight, stirring it every so often to help the oil come out.

### 3. Filtration

**Filtering**: To separate the liquid (which contains the 5-MeO-DMT salt) from the plant matter, filter the mixture through filter paper or a coffee filter. Put the filtered liquid into a clean glass jar.

### 4. Forming the base

- **adding a base,** like sodium hydroxide, slowly to the acidic liquid while keeping an eye on the pH. You want the pH to be between 11 and 12 to turn the 5-MeO-DMT salt back into its freebase form.

- **Mixing:** Make sure the solution is completely changed by stirring it well. This step is very important for getting the 5-MeO-DMT out into a non-polar liquid in the next step.

## 5. Using a non-polar solvent to extract

- **add a solvent**, pour a non-polar solvent into the base solution. Examples of such solvents are naphtha and heptane. The liquid will break down the 5-MeO-DMT freebase.
- **Shaking:** Put the lid on the container and shake it hard for a few minutes to make sure the ingredients are well mixed and the oil is extracted.

- **Separation**: Let the mixture separate into two layers: the upper layer contains 5-MeO-DMT and is a solvent, and the bottom layer is water. The liquid layer should be carefully split up with a separation funnel and moved to a clean container.

## 6. Evaporation

- **To dry the liquid,** pour it into a dish for evaporation. For slow evaporation, use light heat (like a hot plate) or put the dish somewhere with good air flow. After this, rough 5-MeO-DMT crystals are left behind.

- **Optional Recrystallization:** To make the crystals even cleaner, dissolve them in a small amount of warm liquid and let them crystallize again as the solvent cools.

### 7. Stored

- **Dried and stored:** Once the 5-MeO-DMT crystals are completely dry, put them in a clean, sealed container that is out of the way of light and water. If you need to stop wetness from absorbing, use desiccant packs.

### Safety and Legal Things to Think About:

- **Safety:** During the extraction process, wear the right safety gear, like gloves and masks. To keep from breathing in chemical vapors, work in a room with good air flow or under a fume hood.

- **Legal:** Make sure you follow the local rules about getting 5-MeO-DMT and any materials that can be used to make it.

If you carefully and responsibly follow these steps, you can get 5-MeO-DMT from plants and clean it up for study or personal use. When you do extraction work, you should always put safety, law, and ethics first.

## Chapter seven

## Extraction of 5-MeO-DMT from toad venom

To get 5-MeO-DMT out of toad venom, especially from the Colorado River Toad (Bufo alvarius), the venom has to be collected and then processed to separate the compound. The following steps will show you how to get 5-MeO-DMT out of toad venom:

**materials you'll Need:**

There are Colorado River Toads (Bufo alvarius).

- Mirror or glass plate

- A scraper, like a razor blade
- A solvent, like ethanol or methanol
- Dish for evaporation
- Glassware, like flasks and beakers
- pH gauge or pH strips
- Safety gear (goggles, gloves, and a smoke hood)

**How to do it:**

**1. Collation of toad venom**

- **Find Toads:** Usually, you can find Colorado River Toads in the southwestern United States and northern Mexico, especially in the Sonoran Desert.
- **milking procedure,** gently squeeze the toad's parotoid glands to get the poison out and

onto a clean mirror or glass plate. Be careful not to hurt the toad, and give the poison time to dry fully. It could take anywhere from a few hours to a day for this to dry.

## 2. Getting and scraping

To carefully scrape the dried venom off the glass plate or mirror, use a clean razor blade or something similar. Put the poison that you scraped into a clean glass jar.

## 3. Extraction with a solvent

• **Dissolve in Solvent**: Mix with a polar solvent, like ethanol or methanol, and add a small amount to the venom that you have gathered. The solvent will remove the 5-MeO-DMT and any other parts of the venom that can be dissolved.

• **Mixing:** Gently stir the mixture to make sure that all of the venom components are dissolved in the solvent.

### 4. Filtration

• Filtering: Put the solvent-venom mix through filter paper or a coffee filter to separate the liquid (5-MeO-DMT that has been dissolved) from any solid residue or material that has not been dissolved. Put the liquid that has been drained into a clean glass jar.

### 5. Evaporation

• **Evaporation:** Put the solvent-venom mix that has been filtered into a dish for evaporation. For slow evaporation, use light heat (like a hot plate) or put the dish somewhere with good air flow. The process will leave behind rough 5-MeO-DMT crystals or waste.

## 6. Purification

- **Recrystallization:** To make the crystals even cleaner, dissolve them in a small amount of warm liquid (like ethanol) and let them grow back into crystals as the solvent cools. The 5-MeO-DMT crystals are cleaned up in this step.

## 7. Stored

- **Dried and stored:** Once the 5-

MeO-DMT crystals are completely dry, put them in a clean, sealed container that is out of the way of light and water. If you need to stop wetness from absorbing, use desiccant packs.

**Safety and Legal Things to Think About:**

• **Safety:** During the extraction process, wear the right safety gear, like gloves and goggles, to keep from coming into touch with the venom and solvent vapors. Work in a room with good air flow or under a fume hood.

• **Legal:** Make sure you follow the local rules about getting 5-MeO-DMT and any materials that can be used to make it. In many places, it may be against the law to handle Colorado River Toads or get their poison out.

You can get 5-MeO-DMT from toad venom and clean it up for study or personal use by carefully following these steps. When you do extraction work, you should always put safety, law, and ethics first.

## Chapter eight

### Safe ways to handle 5-MeO-DMT

To avoid infection and keep yourself safe, it is very important to handle 5-MeO-DMT safely. Here are some important rules to follow:

**1. Safety gear for individuals (PPE):**

- **Gloves:** To keep from coming into close contact with 5-MeO-DMT, wear disposable gloves made of nitrile or latex.

- **Goggles or safety glasses:** These will protect your eyes from splashes or fumes.

- **Lab coat or protective clothing:** You should cover your skin with the right clothes to lower your risk of exposure.

**2.Taking care of things in a controlled setting:**

- **Fume Hood:** If you're working with powdered forms or liquids that could give off fumes, do the work under a fume hood to make sure there is enough air flow.

- **Well-Ventilated Area:** To avoid breathing in too many vapors or aerosols, work in a room with good air flow.

**3. How to Avoid Contamination:**

**Clean Surfaces:** Make sure that any surface that will come into contact with 5-MeO-DMT is clean and free of any other poisons or dirt.

- **Use Different Tools:** To keep 5-MeO-DMT from getting contaminated with other substances, only use special tools (like glasses and dishes) for handling it.

**4. Putting away and labeling:**

- **Containers That Keep Air Out:** To keep 5-MeO-DMT safe from air and moisture, store it in containers that keep air out.

- **Desiccant Packs:** Put desiccant packs in storage cases to keep the humidity low and stop things from breaking down.

- **Correct Labeling:** Make sure

that packages are clearly marked with what's inside, the concentration (if needed), and the date that it was made to make sure that it is safe to handle and store.

**5. Being ready for an emergency:**

● **Spill Kit:** In case of an accident, keep a spill kit close by. It should have absorbent materials, gloves, and a trash bag.

● **First Aid:** Know what to do in case of an exposure event and have a first aid kit on hand. Get medical help right away if you are exposed or if you start to feel sick.

**6. Disposal:**

- **Proper Disposal:** Get rid of contaminated materials and 5-MeO-DMT that hasn't been used or has expired according to local rules and guidelines for getting rid of hazardous trash.

**7. Training and Being Aware:**

- **Education:** Make sure that anyone who handles 5-MeO-DMT has been taught in safe handling techniques and is aware of the possible dangers of the compound.
- **Risk Assessment:** Before dealing, do a risk assessment to find possible dangers and take the right safety steps.

## What the law says about 5-MeO-DMT

5-MeO-DMT is mostly banned everywhere in the world. It is categorized with other strong psychoactive drugs that are easy to abuse and don't have any known medical uses. Because it is a Schedule I or equivalent substance, it is usually illegal to handle it in any way, including making, buying, having, or giving it to other people, unless it is for research reasons. Before trying to get or use 5-MeO-DMT, users should be fully aware of the legal risks and possible fines in their own countries.

www.ingramcontent.com/pod-product-compliance
Lightning Source LLC
Chambersburg PA
CBHW072001210526
45479CB00003B/1024